CW00470931

ır

s

Vilas Parmar

Practicals of Cell Biology & Genetics

LAP LAMBERT Academic Publishing

Imprint
Any brand names and product names mentioned in this book are subject to trademark, brand or patent protection and are trademarks or registered trademarks of their respective holders. The use of brand names, product names, common names, trade names, product descriptions etc. even without a particular marking in this work is in no way to be construed to mean that such names may be regarded as unrestricted in respect of trademark and brand protection legislation and could thus be used by anyone.

Cover image: www.ingimage.com

Publisher:
LAP LAMBERT Academic Publishing
is a trademark of
International Book Market Service Ltd., member of OmniScriptum Publishing Group
17 Meldrum Street, Beau Bassin 71504, Mauritius

Printed at: see last page
ISBN: 978-613-9-96309-6

Copyright © Vilas Parmar
Copyright © 2018 International Book Market Service Ltd., member of OmniScriptum Publishing Group

Practicals of Cell Biology & Genetics

INDEX

1. Study of various cell organelles

[A] PLASTIDS

Plastids are large cytoplasmic organelles. Plastids are major organelles found in the cells of plants and algae. Plastids are the site of manufacture and storage of important chemical compounds used by the cell. Plastids often contain pigments used in photosynthesis, and the types of pigments present can change or set the cell's color. The term plastid was derived from the Greek word "plastikas" meaning formed or shaped. This term was coined by Schimper in 1885.

In plants, plastids may differentiate into several forms, depending upon which function they need to play in the cell. The plastids are generally separated into two main types, namely chromoplasts and leucoplasts.

[1] Chromoplast:

These are colored plastids (chromo=color; plast=living). They hold in various pigments. They synthesize food materials by photosynthesis. They contain yellow, orange and or red pigments. Chromoplasts are found commonly in flowers and fruits. Chromoplasts also divided into three cases based on their color, namely chloroplast, phaeoplast and rhodoplast.

    (i)   **Chloroplast:** It is in green color. It contains chlorophyll pigments. It is found in higher plants and green algae.

    (ii)   **Phaeoplast:** It is dark brown in color. It contains fucoxanthin pigments. It is found in brown algae, diatoms and dinoflagelates.

    (iii)   **Rhodoplast:** It is reddish in gloss. It contains phycoerythrin. It is base in red algae.

[2] Leucoplasts:

They are non-pigmented plastids (Leuco=white; plast=living). Their main function is to store food materials. They do not involve in synthetic activities. The leucoplasts are subdivided into three types, namely amlyoplast, elaioplast and proteinoplast.

    (i)   **Amlyloplast:** It stores starch and found in tubers, cotyledons and endosperm

    (ii)   **Elaioplast:** It stores oil and found in the epidermal cells.

    (iii)   **Proteinoplast:** It stores protein and found in seeds and nuts.

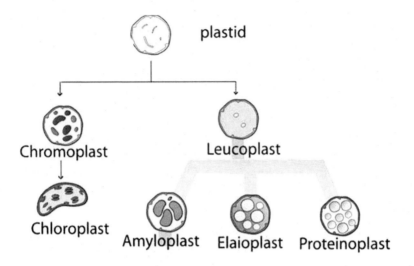

Figure 1: Types of plastids
web ref: 1

[B] CHLOROPLASTS

Chloroplasts are organelles found in plant cells and other eukaryotic organisms that conduct photosynthesis. The word chloroplast is derived from the Greek words *chloros*, which means green, and *plast*, which means form or entity. Chloroplasts are members of a class of organelles known as plastids.

Shape:
Chloroplast varies in shape. They are spheroid or ovoid or discoid in higher plants. They are cup-shaped in chlamydomonas and spirally coiled in spirogyra.

Size:
The size of the plastids varies from species to species. But the size remains constant for a given cell type. In higher plants, it is 4-5microns in length and 1-3microns in thickness. Generally chloroplasts of plants growing in shady places are bigger in size.

Number:
The number of chloroplasts varies from plant to plant, but it remains constant for a given plant. In higher plants there are 20 to 40 chloroplasts per cell or up to 1000 chloroplasts

Structure:

Plant chloroplasts are large organelles (5 to 10 μm long) that, like mitochondria, are bordered by a double membrane called the chloroplast envelope. In summation to the interior and outer tissue layers of the envelope, chloroplasts have a third internal membrane system, called the thylakoid membrane. The thylakoid membrane forms a network of flattened discs called thylakoids, which are often arranged in stacks called grana. Grana are interconnected by branching membraneous tubules called frets (stromal lamellae). Because of this three-membrane structure, the interior organization of chloroplasts is more complex than that of mitochondria. In particular, their three membranes divide chloroplasts into three distinct internal compartments: (1) the intermembrane space between the two membranes of the chloroplast envelope; (2) the stroma, which lies inside the envelope but outside the thylakoid membrane; and (3) the thylakoid lumen.

A thylakoid has a flattened disk form. Within it is an empty field called the thylakoid space or lumen. Photosynthesis takes place on the thylakoid membrane; as in mitochondrial oxidative phosphorylation, it requires the coupling of cross-membrane fluxes with biosynthesis via the dissolution of a proton electrochemical gradient. Planted in the thylakoid membrane is antenna complexes, each of which consists of the light-absorbing pigments, including chlorophyll and carotenoids, as well as proteins that hold the pigments. These complexes are called as quantosomes. This complex, both increases the open area for light capture, and allows capture of photons with a broader range of wavelengths. The energy of the incident photons is absorbed by the pigments and funneled to the reaction centre of this complex through resonance energy transfer. Two chlorophyll molecules are then ionized, producing an excited electron, which then goes along to the photochemical reaction center.

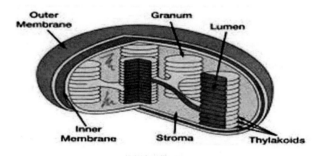

Figure 2: Ultrastructure of Chloroplast

web ref: 2

[C] MITOCHONDRIA

4

Mitochondria are known as energy currency or a powerhouse of the cell. Mitochondria can be seen with a special microscope called electron microscope. They are sac-like structures present in the cytoplasm of the mobile phones. They may be of various shapes threadlike, spherical. Mitochondria have two compartments - an interior compartment and an outer compartment. The substance in the inner compartment is called matrix. The matrix is surrounded by a membrane called inner membrane of mitochondria. The internal membrane is thrown into several folds called cristae. The cristae extended into the matrix. The distance between the folds is continuous with the outer compartment. The outer compartment is surrounded by another membrane - the outer membrane. The outer membrane is smooth and has no projections. The inner membrane, the matrix and the elementary particles in the mitochondria have a big number of enzymes and other proteins required for the breathing and energy output.

Mitochondria are small, frequently between 0.75 and 3 micrometers and are not visible below the microscope unless they are defiled. Unlike other organelles (miniature organs within the cubicle), they possess two membranes, an outer one and an internal one. Each membrane has different functions. Mitochondria are split into different compartments or regions, each of which carries out distinct roles.

Some of the major regions include the:

(i) **Outer membrane:** Small molecules can pass freely through the outer membrane. This outer portion includes proteins called porins, which form channels that allow proteins to cross. The outer membrane also hosts a number of enzymes with a broad variety of uses.

(ii) **Intermembrane space:** This is the area between the inner and outer membranes.

(iii) **Inner membrane:** This membrane holds proteins that have several roles. Because there are no porins in the inner membrane, it is impermeable to most molecules. Atoms can only traverse the inner membrane in special membrane transporters. The inner membrane is where most ATP is created.

(iv) **Cristae:** These are the folds of the inner membrane. They increase the open area of the membrane, hence increasing the space available for chemical reactions.

(v) **Matrix:** This is the space within the inner membrane. Containing hundreds of enzymes, it is important in the production of ATP. Mitochondrial DNA is housed here.

Different cell types have different numbers of mitochondria. For example, mature red blood cells have none at all, whereas liver cells can hold more than 2,000. Cells with a high requirement for energy tend to receive larger numbers of mitochondria. Approximately 40 percent of the cytoplasm of heart muscle cells is carried up by mitochondria. Although mitochondria are often described as an ellipse-shaped organelles, they are constantly dividing (fission) and bonding together (fusion). Thus, in reality, these organelles are linked together in ever-changing networks.

5

Likewise, in sperm cells, the mitochondria are spirals in the mouthpiece and provide energy for tail motion.

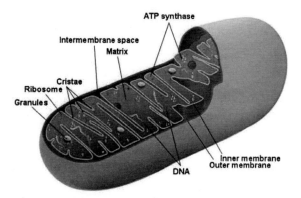

Figure 3: Ultrastructure of Mitochondria

web ref: 3

[D] CELL WALL

The cell wall is distinguished into three parts- middle lamella, primary wall and secondary wall.

(1) Middle lamella: It is a thin amorphous cementing layer between two adjacent cells. It is the first layer, which is deposited at the time of cytokinesis. It is composed of calcium and magnesium pectate.

(2) Primary wall: It is usually thin, elastic and extensible in growing cells. It grows by the addition of wall material inside the surviving one. Such growth is termed as inter susception. This wall consist of a loose network of cellulose microfibrils. These microfibrils in fungi are made up of a polymer of acetyl Glucosamine. The matrix of the wall is mainly composed of water, hemicellulose, pectins and glycoproteins.

(3) Secondary wall: After maturity, a thick secondary wall is laid inner in the primary wall by accretion or deposition of materials. It is set up in many strata.

It is the outer rigid protective supportive and semi transparent covering of plant cells, fungi and some protists. Cell wall was first considered in the cork cells by Hooke in 1665. Its thickness varies in different types of cells from $0.1\ \mu m$ to $10\ \mu m$. The cell wall is a non-living extracellular

6

secretion or matrix of the cell which is nearly pressed to it. It is metabolically active and is capable of growth.

Roles of Cell Wall:
(i) Protects the protoplasm against mechanical injury,
(ii) Protects the cell from an attack of pathogens,
(iii) Provides rigidity and shape to the cell,
(iv) Counteracts osmotic pressure.
(v) Gives strength to the land plants to withstand gravitational forces,
(vi) By its growth the wall helps in cell expansion,
(vii) Pits present in the wall help produce a protoplasmic continuum or simplest amongst cells,
(viii) Walls prevent bursting of plant cells by inhibiting excessive endosmosis.
(ix) The wall has some enzymatic activity associated with metabolism,
(x) In many cases, the wall takes part in offense and defense,
(xi) Cutin and suberin of the cell wall reduce the loss of water through transpiration,
(xii) Walls of sieve tubes, tracheids and vessels are specialized for long distance shipping,
(xiii) Some seeds store food in the form of hemicellulose in the cell wall.

Chemical Composition of Cell Wall:
1. Matrix:
Water— 60%, Hemicellulose— 5- 15%, Pectic Substances-2-8%, Lipids-0.5-3.0%, Proteins—1-2%.

2. Micro fibrils:
Cellulose/fungus cellulose— 10-15%.

3. Other Ingredients:
Lignin, cutin, suberin, silica (silicon dioxide), minerals (e.g., iron, calcium, carbonate), waxes, tannins, resins, gum— variable.

The cell wall structure:
A cell wall can give birth up to three parts— middle lamella, primary wall and secondary wall.

Middle Lamella:
It is a thin, amorphous and cementing layer between two neighboring cells. Middle lamella is the first layer which is lodged at the time of cytokinesis. It is just like brick work of the common wall between two adjacent rooms.

/

Middle lamella is absent on the outer side of surface cells. It is made up of calcium and magnesium pectates. The softening of ripe fruits is caused by partial solubilisation of pectic compounds to produce jelly like consistency.

Primary Wall:
It is the first formed wall of the cell which is produced inner to the middle lamella. The primary wall is commonly thin (0.1-3.0 µm) and capable of extension. It grows by intussusceptions or addition of fabrics within the existing wall. Some cells have only primary wall, e.g., Leaf cells, fruit cells, of cortex and pith.

Primary wall consists of a number of micro fibrils embedded in the amorphous gel like matrix or ground substance. In the majority of plants, the micro fibrils are formed of cellulose. They are synthesized at the plasma membrane by particle rosettes (terminal complexes) having cellulose synthetize enzyme.

The wall is organized of a polymer of P, 1-4 acetyl Glucosamine or fungus cellulose in many fungi. Fungus cellulose is similar to chitin present in the exoskeleton of insects. Micro fibrils are oriented variously according to the shape and thickening of the wall. Usually they are put in a loose network due to incomplete cross-linking.

The matrix of the wall consists of water, pectin, hemicelluloses and glycoproteins. Pectin is the filler substance of the ground substance. Proteins are structural and enzymatic. Protein expansion is involved in loosening and expansion of cell wall through incorporation of more cellulose. Hemicellulose binds micro fibrils with matrix.

Secondary Wall:
It is produced in some mature cells when the latter have stopped growth, e.g., tracheids, vessel elements, fibers, collenchyma's. Secondary wall is laid inner in the primary wall by accretion or deposition of materials over the surface of existing building. It is thick (3—10 µm) and made up of at least three layers, sometimes more (e.g., Latex tube of Euphorbia Milli). They are identified as S1, S2, S3, Sx, etc.
The innermost layer of the secondary wall is sometimes distinct both chemically as well as in staining properties due to the presence of xylem. It is then called tertiary wall, e.g., Tension wood in gymnosperms. Secondary wall may be absent, irregularly deposited or formed uniformly in the xylem. This results in differentiation of cells— parenchyma, collenchyma, sclerenchyma, tracheids and vessels.

8

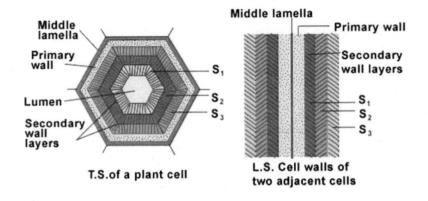

Figure 4: Structure of Cell wall

web ref: 4

Structure: Golgi body is seen in the form of three components.

1. Cisternae: These are tubular, flat, fluid filled sacs. They show 200 to $300A^0$ width. Each pocket is hidden by two membranes. In a dictyosome 3 to 7 cisterne are present. They are all piled one above the other. Their convex side is towards the nucleus and their concave surface is towards plasma membrane. The convex side of the cisternae is called forming face. The concave surface is called maturing face. It shows big secreting vesicles. These secretory vesicles store secretory substances. They may develop into lysosomes.

Polarity of cisternae: The cisternae shows maturing face and forming face. Forming face is convex and towards the nucleus. The smooth E.R. gives vesicles. They combine to form cisternae.

2. Golgi vesicles: On the forming face of Golgi cisternae small vesicles are present. They are 400 A^0 width. They usually develop from E.R.

3) Secretory vesicles: On the maturing face of golgi cisternae secretory vesicles are present. They contain secretory products of golgi. They finally change into lysosomes.

9

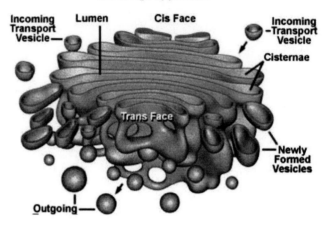

The Golgi Apparatus

Incoming Transport Vesicle —

Lumen

Cis Face

Incoming -Transport Vesicle

Cisternae

Trans Face

— Newly Formed Vesicles

Outgoing —

Figure 5: Ultrastructure of Golgiapparatus

web ref: 5

[F] LYSOSOMES

Established along the cell structure and types of cell organelles present, living organisms has been separated into two categories, prokaryotes, and eukaryotes. Eukaryotes are multicellular organisms. Prokaryotes are generally unicellular organisms; however, few multicellular organisms are prokaryotes. A prokaryotic cell does not contain well-defined membrane-bound organelles. Most of the organelles which a eukaryotic cell contains are absent in a prokaryotic cell. The lysosome is one among them.

Lysosomes are small membrane-bound pocket-like structures which release digestive enzymes that break down food. They work as a waste bin on the cell and prevent the cell clean. Lysosomes are present in eukaryotic cells but a prokaryotic cell lacks them. When a foreign matter gets into the cell, they release their enzymes which split the strange substance into tiny bits and defeat them. They also remove the old and damaged or dead organelles of the cell and thus, protect the cell from further damages and issues. Lysosomes are also responsible for the digestion of food which we deplete. They release an enzyme which is potent enough to snap off any organic issue. Rough endoplasmic reticulum (RER) in the cell is responsible for the synthesis of these enzymes.

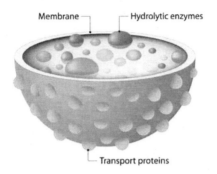

Membrane ⌐ Hydrolytic enzymes

⌐ Transport proteins

Figure 6: Ultrastructure of Lysosome

web ref: 6

A damaged or infected cell fails to perform its functions due to metabolic irregularities or other causes. In such places, the lysosomes inside a cell burst out and engulf their own mobile phone. So, it protects the other cells of the body from further infections and damages. Since lysosomes engulf their own cell, they are recognized as "suicide bags" of a cell. Hence, it delivers a big part in immunity of the consistency.

[G] PEROXISOMES

A type of organelle found in both animal cells and plant cells, a peroxisome is a membrane-bound cellular organelle that contains mostly enzymes. Peroxisomes perform important functions, including lipid metabolism and chemical detoxification. They too carry out oxidation reactions that break down fatty acids and amino acids.

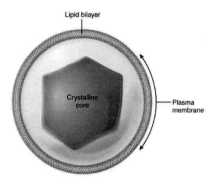

Figure 7: Ultrastructure of Peroxisome   web ref: 7

Peroxisomes are membrane-bound organelles that carry an abundance of enzymes for detoxifying harmful substances and lipid metabolism.

In contrast to the digestive enzymes found in lysosomes, the enzymes within peroxisomes serve to remove hydrogen atoms from various molecules to oxygen, producing hydrogen peroxide (H_2O_2). In this way, peroxisomes neutralize poisons, such as alcohol, that enter the body. In order to value the importance of peroxisomes, it is necessary to interpret the concept of reactive oxygen species.

Reactive oxygen species (ROS), such as peroxides and free radicals, are the highly-reactive products of many normal cellular processes, letting in the mitochondrial reactions that make ATP and oxygen metabolism. Examples of ROS include the hydroxyl radical OH, H_2O_2, and superoxide (O^{-2}). Some ROS are important for certain cellular functions, such as cell signaling processes and immune responses against foreign substances. Many ROS, however, are harmful to the body. Free radicals are reactive because they contain free, unpaired electrons; they can easily oxidize other molecules throughout the cell, causing cellular damage and even cell death. Free radicals are believed to play a part in many destructive processes in the body, from cancer to coronary artery disease.

Peroxisomes averse reactions that neutralize free radicals. They make big quantities of the toxic H_2O_2 in the process, but contain enzymes that convert H_2O_2 into water and oxygen. These byproducts are then safely discharged into the cytoplasm. Like miniature sewage treatment plants, peroxisomes neutralize harmful toxins so that they do not cause harm to the cell. The liver is the organ chiefly responsible for detoxifying the blood before it moves throughout the body; liver cells contain an exceptionally high number of peroxisomes.

2. Study the various stages of cell divisions (Mitosis/Meiosis) in plant cells

[A] Mitosis:

Essentials:

Microscope, microscope slide, cover slips, prepared slides of onion root tip mitosis, Acetocarmine stain in dropping bottle, 1 M hydrochloric acid (HCL) in dropping bottle, Watch glasses, Dissecting needles, Scalpel or razor blades, Forceps, Alcohol lamp or slide warmer, Supply of onion bulbs Beakers in which to let onions sprout, caching solution.

Method of slide preparation:

1) Clip the terminal 1 cm of the root crown from a growing onion bulb and apply it right away.
2) Put a few ml of 1 M HCL in a watch glass enough to cover root tip. *(Caution! HCl is caustic! Handle with care, and flush well with clean water if you get any on your skin or clothing.)*
3) Into this acid place the terminal 3 or 4 mm of the 1-cm-long onion root.
4) In a short time (a few minutes) the root crown will feel soft when touched with a dissecting needle.
5) Now, using forceps or a needle, pick up the softened root tip and transfer it to a drop of acetocarmine stain on a clean slide.
6) Using a razor blade or a sharp scalpel, chop the root tip into tiny bits. Note: Iron in the scalpel or dissecting needle reacts with the acetocarmine stain *(Caution! Acetocarmine is caustic and corrosive! Handle with care, and flush well with clean water if you get any on your skin or clothing. Acetocarmine may stain skin and permanently stain clothing.)*, to make a better staining reaction.
7) At one time this operation is complete, apply a clean cover glass to the slide and heat it gently over an alcohol lamp or slide warmer. **Do not boil it.** Then reversed the slide on a paper towel and push downward firmly, applying pressure with your thumb over the back glass. This should flatten the cells and spread them so they can be kept under the microscope.
8) Examine under lower power (100x) and then under high power (430x).

Conclusion:

The slide shows almost all the stage of mitosis

The stages of active mitosis are given below:

A. Interphase

The interphace is the cell cycle stage in which cells spend the bulk of their time. An interphase cell is busy undergoing the chemical responses that facilitate energy transfer, growing, and preparing to divide. The interface is not part of mitosis, which is set as active cell division. Instead, it is the living stage of the cell. The chromosomes cannot be distinctly ascertained, as they are in the diffuse form known as chromatin. The factors on the chromosomes are actively being transcribed and translated, though, which genes are active depends on the individuality and role of the single cell. The stages known as G1, S, and G2 occur during intervals, and are required in normal cell growth (G1), DNA synthesis (S), and protein and microtubule synthesis preceding mitosis (G2).

B. Prophase

Prophase is the beginning phase of mitosis. The chromosomes condense and get visible. Although they have already duplicated, the chromosomes are not yet visible as separate sister chromatids. The nuclear membrane breaks down, and the spindle fibers begin to form at opposite poles of the cell.

C. Metaphase

In this second phase of mitosis, the duplicated chromosomes unwind from each other, becoming visible as two identical **sister chromatids** attached only at a slightly constricted region, the **centromere**. It is only now that the chromosomes take on the "X" shape so often seen in illustrations. Only this "X" actually represents not one, but two chromosomes: the identical sister chromatids produced during DNA replication. Spindle fibers attach to the kinetochore, a protein structure located at the centromere, and the chromosomes are arranged on the equator of the mobile phone.

D. Anaphase

Spindle fibers shorten, the kinetochores separate, and the chromatids (daughter chromosomes) are pulled apart and begin moving to opposite poles.

E. Telophase

The daughter chromosomes arrive at the poles and the spindle fibers disappear.

Cytokinesis:

In plant cells, wall formation starts in the middle of the cell and grows outward to match the existing lateral walls. The establishment of the new cell wall begins with the establishment of a simple precursor, called the cell-plate that represents the middle lamella between the walls of two contiguous cells. At the time of cytoplasmic division, organelles like mitochondria and plastids get distributed between the two daughter cells.

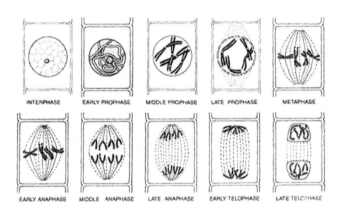

Figure 8: Different stages of Mitosis in plant cell

web ref: 8

[B] Meiosis:

The specialized kind of cell division that cuts down the chromosome number by half results in the production of haploid daughter cells. This sort of class is called reduction division.

Essentials:

Flower buds, Microscope, microscope slide, cover slip, prepared slides of onion root tip mitosis, Acetocarmine stain in dropping bottle, 0.1 N hydrochloric acid (HCl) in dropping, spirit lamp.

Process:

15

1) Open a flower bud and carefully take out few antheridia on to the clean slide.
2) Put a drop of 0.1N HCl and heat repeatedly over spirit lamp about 2-5 sec.
3) Place a few drops of acetocarmine stain and allow to stain for 10-15 minutes.
4) Use a clean cover glass to the slide and heat it gently over an alcohol lamp or slide warmer. **Do not boil it.** Then reversed the slide on a paper towel and push downward firmly, applying pressure with your thumb over the back glass. This should flatten the cells and spread them so they can be kept under the microscope.
5) Examine under lower power (100x) and then under high power (430x).

Observation:

Following stages can be realized in different slide of meiosis.

Meiosis involves two sequential cycles of nuclear and cell division called meiosis I and meiosis II, but only a single round of DNA reproduction. Meiosis I is initiated after the parental chromosomes have replicated to produce identical sister chromatids at the S phase. Meiosis involves pairing of homologous chromosomes and recombination between them. Four haploid cells are made at the final stage of meiosis II. Meiotic events can be grouped under the succeeding phases:

Meiosis I	Meiosis II
Prophase I	Prophase II
Metaphase I	Metaphase II
Anaphase I	Anaphase II
Telophase I	Telophase II

[1] Meiosis I

[i] Prophase I:

Prophase of the first meiotic division is typically longer and more complex when compared to prophase of mitosis. It has been further subdivided into the following five stages based on chromosomal behavior, i.e., Leptotene, Zygotene, Pachytene, Diplotene and Diakinesis. During leptotene stage the chromosomes become gradually visible under the light microscope. The compaction of chromosomes continues throughout leptotene. This is accompanied by the second stage of prophase I called zygotene. During this stage chromosomes start pairing together and this operation of connection is called Synapsis. Such paired chromosomes are called homologous chromosomes. Electron micrographs of this level suggest that chromosome Synapsis is accompanied by the establishment of complex structure called synaptonemal complex. The complex formed by a couple of synapses homologous chromosomes is called a bivalent or a foursome. However, these are more distinctly visible on the adjacent level. The first two stages of prophase I are relatively short-lived compared to the next stage that is pachytene. During this stage bivalent chromosomes now clearly appears as tetrads. This phase is characterized by the appearance of recombination nodules, the situations at which crossing over occurs between non-sister chromatids of the homologous chromosomes. Crossing over is the interchange of genetic material between two homologous chromosomes. Crossing over is also an enzyme-mediated process and the enzyme involved is called recombinase. Crossing over leads to recombination of genetic material on the two chromosomes. Recombination between homologous chromosomes is completed by the end of pachytene, leaving the chromosomes linked at the sites of crossing over. The beginning of diplotene is recognised by the dissolution of the synaptonemal complex and the tendency of the recombined homologous chromosomes of the bivalents to separate from each other except at the sites of crossovers. These X-shaped structures, are called chiasmata. In oocytes of some vertebrates, diplotene can last for months or years. The final stage of meiotic prophase I is diakinesis. This is marked by terminalisation of chiasmata. During this phase the chromosomes are fully digested and the meiotic spindle is set up to prepare the homologous chromosomes for separation. By the end of diakinesis, the nucleolus disappears and the nuclear envelope also breaks down. Diakinesis represents transition to metaphase.

[ii] **Metaphase I:** The bivalent chromosomes align at the equatorial plate. The microtubules from the opposite poles of the spindle attach to the pair of homologous chromosomes.

[iii] **Anaphase I:** The homologous chromosomes separate, while sister chromatids remain associated at their centromeres.

[iv] **Telophase I:** The nuclear membrane and nucleolus reappear, cytokinesis follows and this is called as diad of cells. Although in many cases the chromosomes do undergo some dispersion, they do not achieve the extremely extended state of the interface nucleus. The stage between the two meiotic divisions is called interkinesis and is more often than not short lived. Interkinesis is followed by prophase II, a much simpler prophase than prophase I.

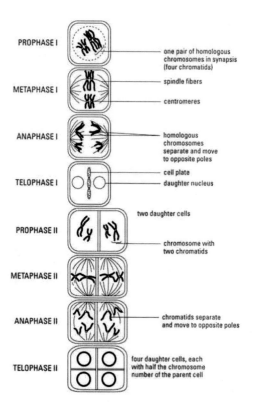

PROPHASE I		one pair of homologous chromosomes in synapsis (four chromatids)
METAPHASE I		spindle fibers
		centromeres
ANAPHASE I		homologous chromosomes separate and move to opposite poles
TELOPHASE I		cell plate
		daughter nucleus
		two daughter cells
PROPHASE II		chromosome with two chromatids
METAPHASE II		
ANAPHASE II		chromatids separate and move to opposite poles
TELOPHASE II		four daughter cells, each with half the chromosome number of the parent cell

Figure 9: Different stages of Mitosis in plant cell

Web ref: 9

[2] Meiosis II:

[i] **Prophase II:** Meiosis II is initiated immediately after cytokinesis, usually before the chromosomes have fully elongated. In contrast to meiosis I, meiosis II resembles a normal mitosis. The nuclear membrane disappears by the final stage of prophase II. The chromosomes again become compact.

[ii] **Metaphase II:** At this stage the chromosomes align at the equator and the microtubules from opposite poles of the spindle get attached to the kinetochores of sister chromatids.

[iii] **Anaphase II:** It begins with the simultaneous splitting of the centromere of each chromosome (which was carrying the sister chromatids together), permitting them to move toward opposite poles of the cell.

[iv] **Telophase II:** Meiosis end with telophase II, in which the two groups of chromosomes once again get enclosed by a nuclear envelope; cytokinesis follows resulting in the establishment of the tetrad of cells i.e., Four haploid daughter cells are obtained.

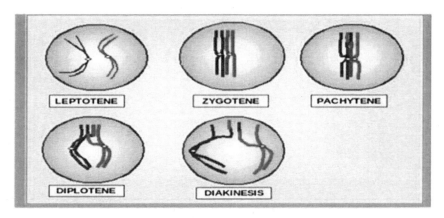

Figure 10: Different stages of Prophase-1 in plant cell

Web ref: 10

3. Study of Giant chromosomes from Salivary glands of Drosophila (Polytene),Chyronomous larva (Lamp brush)

Essentials:

Dissecting scope (one per Lab Group) 2) Bottle of wild type Drosophila stock 3) 10 ml Saline (0.7% NaCl) 4) 1 ml 45% Acetic Acid 5) 0.5 ml Orcein stain 6) Kimwipes 7) H_2O wash bottle 8) In Lab Bench Box: - Microscope slides - Coverslips - 2 pairs of Forceps

Salivary Gland Dissections

1. Use your forceps to remove 8-10 actively crawling larvae from the sides of the cultivation bottle. Put them on a dry microscope slide (slightly to one side).

2. Use a Pasteur pipette to place a drop of saline solution in the heart of the same microscope slide. Use your forceps to remove one of the larvae to the saline drop, with its tracheal tubes facing up.

3. Use one pair of forceps to hold the anterior end of the larva in place at the point just above the neural ganglion in Figures 3 and 4, while using a second pair of forceps to take hold of the tip layer of cuticle from a view on the larva at about one third body length from the anterior end.

4. The salivary glands and ventral ganglion (brain) will usually stay attached to the head area, split from the eternal sleep of the body, after a clean dissection. You may have to use your forceps to search for the salivary glands among the analyzed material. If this bombs, just start over with a new larva. These dissections are difficult to overcome, but reasonable results can be obtained with a small exercise.

5. As you dissect the salivary glands, you can either keep them set aside in a separate saline drop on the slide used for your dissections or transfer them to a separate saline drop on a different coast. In handling the salivary glands, either grab them from their less fragile smallest diameter base or use your forceps to scoop them from underneath.

6. After you have accumulated 5-6 salivary glands, prepare a slide for a squash of 1-2 glands. Use a Kimwipe to wipe its surface white, then burn out any dust from its surface before placing a

10µl drop of Lacto/Aceto Orcein stain on the glide. One partner of the lab group can cover this project while the other (probably the partner who made out the dissection) handles step # 7 below.

7. Then place a drop of 45% Acetic Acid (HOAc) fixative on the coast side by side to the drop of saline holding your salivary glands and use your forceps to remove the salivary glands from the saline to it and hold in place for ~30 seconds. Then immediately transfer the glands to the drop of Lacto/Aceto Orcein stain on the slide from step # 6.

8. Place a clean coverslip (dust blown from its surface) onto the surface of the glands and use your forceps to gently tap the surface of the coverslip in a circular pattern. The coverslip should rotate somewhat freely on the airfoil of the microscope slide; this is needed to burst the membranes of the polytene nuclei and allow the chromosome arms to disperse.

9. Put the slide on your microscope. Using the lowest power objective and standard light source, scan around the slide until you find the squashed material. It is frequently useful to first locate the squashed material on the slide by eye. Then put the slide along the microscope stage with this material centered in the light shaft. This will make it easier to find your chromosomes under the microscope. First locate the focal plane of your chromosomes by adjusting the focus knob until the fluid background material becomes visible. You can then begin to scan around the slide to find the chromosomes.

11. If the chromosomes look sufficiently well spread, you can now invert the slide and coverslip onto a sheet of Kimwipe. Then utilize your thumb to give pressure to the coverslip while using the middle and index fingers of your other hand to secure the place of the slide and coverslip. This squashing action will bring about further opening of the chromosomes and remove excess liquid between the slide and coverslip to allow maximal attachment of the chromosomes to the coast.

12. If the nuclei were found to be still intact, you force out try to tear them if there is evidence of liquid between the slide and coverslip as evidenced by movement of desktop fabric. To rupture the nuclei, use your forceps to tap the surface of the coverslip as described in Steps 8-10. It may be necessary to start over with some newly dissected salivary glands. Preparing polytene chromosome squashes is part science/part art. It requires a lot of exercise and a little luck to capture the quality of squash.

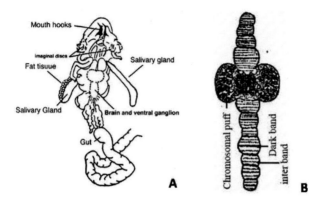

Figure 11: Salivary glands of Drosophila [A]; Polytene chromosome [B]

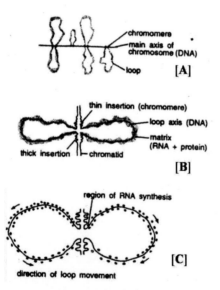

Figure 12: Structure of lampbrush chromosome.
(A) Gross structure of chromosome [a part]; (B) Fine structure of chromosome; (C) RNA synthesis in loop.

4. Induction of polyploidy in onion root tips

Two groups of onion root tips will be used, and a chromosome squash will be performed in order to visualize the state of the cell, in terms of mitosis. The chromosome squash allows the detective to keep cells and several phases of mitosis under a light microscope.

Colchicine is an alkaloid compound derived from the corm and other constituents of the Autumn Crocus, Colchicum autumnale. Colchicine induces the disassembly of microtubule fibers and therefore stops the mitotic process. It is as well recognized as a mitotic poison. If applied right, it can be used to stop mitosis "in its rails" so that chromosome morphology can be studied, chromosome counts can be caused, or induction of polyploidy can be done.

Essentials:
Microscope, microscope slide, cover slips, prepared slides of onion root tip mitosis, Acetocarmine stain in dropping bottle, 1 M hydrochloric acid (HCl) in dropping, Watch glasses, Dissecting needles, Scalpel or razor blades, Forceps, Alcohol lamp or slide warmer, Supply of onion bulbs Beakers in which to let onions sprout, caching solution.

Experiment with design:
One group of onion root cells will be developed in a beaker with plain water under normal conditions (12-hour light/dark cycle, 75OF, etc.). The second group of stem cells will be produced for a short period of time in a dilute colchicine solution.

Method of slide preparation:
1) Clip the terminal 1 cm of the root crown from a growing onion bulb and apply it right away.
2) Put a few ml of 1 M HCl in a watch glass enough to cover root tip. *(Caution! HCl is caustic and corrosive! Handle with care, and flush well with clean water if you get any on your skin or clothing.)*
3) Into this acid place the terminal 3 or 4 mm of the 1-cm-long onion root.
4) In a short time (a few minutes) the root crown will feel soft when touched with a dissecting needle.
5) Now, using forceps or a needle, pick up the softened root tip and transfer it to a drop of acetocarmine stain on a clean slide.
6) Using a razor blade or a sharp scalpel, chop the root tip into tiny bits. Note: Iron in the scalpel or dissecting needle reacts with the acetocarmine stain *(Caution! Acetocarmine is caustic and corrosive! Handle with care, and flush well with clean water if you get any on your skin or clothing. Acetocarmine may stain skin and permanently stain clothing.)*, To make a better staining reaction.
7) At one time this operation is complete, apply a clean cover glass to the slide and heat it gently over an alcohol lamp or slide warmer. **Do not boil it.** Then reversed the slide on a paper towel and push downward firmly, applying pressure with your thumb over the back glass. This should flatten the cells and spread them so they can be kept under the microscope.

8) Examine under lower power (100x) and then under high power (430x).
9) Apply the same routine to examine your treatment and control bases.
10) Calculate the number of cells you can distinguish in each phase of mitosis. Use these data for statistical analysis, to decide whether there is a significant divergence between your treatment and control samples.
11) For comparison, you may care to examine a commercially prepared slide of an onion root tip (available from your teacher). Observe the slide under low and high power.

Conclusion:
We can experience under the microscope cell division without cell wall formation may be affected, leading to double of chromosome number.

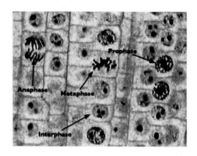

Figure 13: Polyploidy in Onian root tip.

Web ref: 12

24

5. Study of permanent Slides/Charts/Models/Photographs of elements of genetics

[A] MALE STERILITY

Types of male sterility. Male sterility in plants can be controlled by nuclear genes or cytoplasm or by both. Therefore, broadly there are at least three different mechanisms for control of male sterility in plants. These three characters would be briefly discussed in this part.

1. **[1] Genetic male sterility.** In this case, male sterility is controlled by a single gene and is recessive to fertility, and then that the F1 individuals would be rich. In the F2 generation, the productive and sterile individuals will segregate into 3 : 1 ratio.

2. **[2] Cytoplasmic male sterility.** In several crops like maize, cytoplasmic control of male sterility is known. In such cases, if female parent is male sterile, F_1 progeny would always be male sterile, because cytoplasm is mainly derived from egg obtained from male sterile female parents.

3. **[3] Cytoplasmic-genetic male sterility.** In certain other instances, although male sterility is wholly controlled by cytoplasm, but a restorer gene if present in the nucleus will restore fertility. For example, if female parent is male sterile, then genotype (nucleus) of male parent will determine the phenotype of F1 progeny. The male sterile female parent will receive the recessive genotype (RR) with respect to restorer gene. If male parent is *RR*, F_1 progeny would be fertile (*Rr*). On the other hand, if male parent is *rr*, the progeny would be male sterile. If F_1 individual (*Rr*)is testcrossed, 50% fertile and 50% male sterile progeny would be obtained (Fig. 18.14).

Cytoplasmic male sterility in maize. Rhoades in 1933, reported the analysis of first cytoplasmic male sterile plants in maize and demonstrated that male sterility was contributed by female parent and that nuclear genes had no influence. This was established by crossing male sterile plants with a extensive range of fertile males and by keeping that in subsequent generations all progenies were male sterile. In corn, three male sterile sources (cms) are recognized, which are called T, C and S. The normal male fertile cytoplasm is called N cytoplasm. Apiece of the three cms cytoplasms exhibits strict maternal inheritance and yet when all chromosomes were replaced by a male fertile source through backcrosses, male sterility could not be overcome. These cytoplasms were a big asset in the production of hybrid corn seed at commercial scale. It was likewise shown that T (Texas) cytoplasm was associated with (i) susceptibility to two diseases, namely southern corn leaf blight disease (Bipolaris maydis race T, at one time known as Helminthosporium maydis) and yellow leaf blight (Phyllstica maydis); (ii) an unusual mitochondrial gene T-urfl3, which encodes a 13 kilodalton polypeptide (URF13). An interaction between fungal toxins and URF3 accounts for specific susceptibility to the above two fungal pathogens. Due to such an association with diseases, use of T cytoplasm in hybrid seed production was greatly cut short.

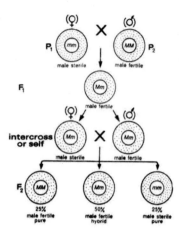

Figure 14: Genetic Male sterility.

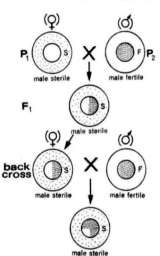

Figure 15: Cytoplasmic Male sterility.

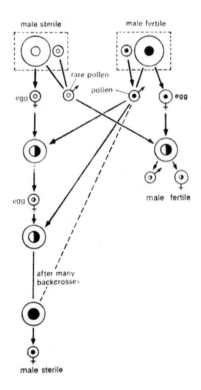

Figure 16: Male sterility in Maize.

Web ref: 13

[B] CIRCULAR GENETIC MAP OF CHLOROPLAST GENOME IN CHLAMYDOMONAS

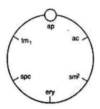

Figure 17: Circular genetic map of chloroplast genome in Chlamydomonas

Web ref: 14

It has been noted that in the dihybrid genetic crosses, a recombination frequency of less than 50% indicates linkage of the two genes involved. Morgan and Sturtevant found that the recombination frequency varies for different mutant gene pairs. The difference in the recombination frequency was found to be associated with the distance between the genes.

The frequency of crossing over between two related genes is directly relative to the distance between them in the chromosome. This is named Morgan and Sturtevants hypothesis.

A chromosome map is a pictorial representation of a linkage group in the shape of a course which shows by points the sequence of the genes and the proportional distance between the genes it controls. A chromosome map is also called a linkage map or genetic map.

Chromosome mapping is established on two under mentioned genetic principles. They are-

1. The genes that are ordered in a linear order along the chromosome.

2. That the frequency of crossing over between two genes is directly relative to the distance between them in the chromosome. Since the frequency of crossing over is used in preparing chromosome maps, the latter is also called crossover maps.

Chromosome map unit:
A frequency of one per cent crossing over between two genes is known as one unit of map distance between these genes and it is termed a Morgan after the name of T. H. Morgan.

[C] CELL CYCLE & CANCEROUS CELL

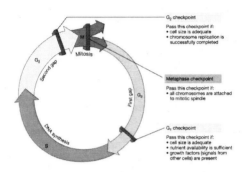

Figure 18: Cell cycle.   Web ref: 15

Cancer is basically a disease of uncontrolled cell division. Its development and progression are usually linked to a series of changes in the activity of cell cycle regulators. For example, inhibitors of the cell cycle keep cells from dividing when conditions aren't right, so too little activity of these inhibitors can promote cancer. Similarly, positive regulators of cell division can lead to cancer if they are too active. In most cases, these changes in activity are due to mutations in the genes that encode cell cycle regulator proteins.

Cell cycle regulators and cancer

Different types of cancer involve different types of mutations, and, each individual tumor has a unique set of genetic alterations. In general, however, mutations of two types of cell cycle regulators may promote the development of cancer: positive regulators may be overactivated (become oncogenic), while negative regulators, also called tumor suppressors, may be inactivated.

Oncogenes

Positive cell cycle regulators may be overactive in cancer. For instance, a growth factor receptor may send signals even when growth factors are not there, or a cyclin may be expressed at abnormally high levels. The overactive (cancer-promoting) forms of these genes are called **oncogenes**, while the normal, not-yet-mutated forms are called **proto-oncogenes**. This naming system reflects that a normal proto-oncogene can turn into an oncogene if it mutates in a way that increases its activity. Mutations that turn proto-oncogenes into oncogenes can take different forms. Some change the amino acid sequence of the protein, altering its shape and trapping it in an "always on" state. Others involve **amplification**, in which a cell gains extra copies of a gene and thus starts making too much protein. In still other cases, an error in DNA repair may attach a proto-oncogene to part of a different gene, producing a "combo" protein with unregulated activity

Many of the proteins that transmit growth factor signals are encoded by proto-oncogenes. Normally, these proteins drive cell cycle progression only when growth factors are available. If one of the proteins becomes overactive due to mutation, however, it may transmit signals even when no growth factor is around. In the diagram above, the growth factor receptor, the Ras protein, and the signaling enzyme Raf are all encoded by proto-oncogenes[11]. Overactive forms of these proteins are often found in cancer cells. For instance, oncogenic Ras mutations are found in about 90% of pancreatic cancers. Ras is a G protein, meaning that it switches back and forth between an inactive form (bound to the small molecule GDP) and an active form (bound to the similar molecule GTP). Cancer-causing mutations often change Ras's structure so that it can no longer switch to its inactive form, or can do so only very slowly, leaving the protein stuck in the "on" state.

Tumor suppressors

Negative regulators of the cell cycle may be less active (or even nonfunctional) in cancer cells. For instance, a protein that halts cell cycle progression in response to DNA damage may no longer sense damage or trigger a response. Genes that normally block cell cycle progression are known as **tumor suppressors**. Tumor suppressors prevent the formation of cancerous tumors when they are working correctly, and tumors may form when they mutate so they no longer work. One of the most important tumor suppressors is **tumor protein p53**, which plays a key role in the cellular response to DNA damage. p53 acts primarily at the G_1 checkpoint (controlling the G_1 to S transition), where it blocks cell cycle progression in response to damaged DNA and other unfavorable conditions[13].

When a cell's DNA is damaged, a sensor protein activates p53, which halts the cell cycle at the G_1 checkpoint by triggering production of a cell cycle inhibitor. This pause buys time for DNA repair, which also depends on p53, whose second job is to activate DNA repair enzymes. If the damage is fixed, p53 will release the cell, allowing it to continue through the cell cycle. If the damage is not fixable, p53 will play its third and final role: triggering apoptosis (programmed cell death) so that damaged DNA is not passed on.

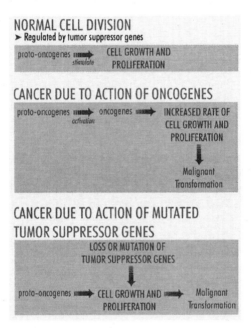

Figure 19: Cancer cell

Web ref: 16

In cancer cells, p53 is often missing, nonfunctional, or less active than normal. For example, many cancerous tumors have a mutant form of p53 that can no longer bind DNA. Since p53 acts by binding to target genes and activating their transcription, the non-binding mutant protein is unable to do its job. When p53 is defective, a cell with damaged DNA may proceed with cell division. The daughter cells of such a division are likely to inherit mutations due to the unrepaired DNA of the mother cell. Over generations, cells with faulty p53 tend to accumulate mutations, some of which may turn proto-oncogenes to oncogenes or inactivate other tumor suppressors. p53 is the gene most commonly mutated in human cancers, and cancer cells without p53 mutations likely inactivate p53 through other mechanisms (e.g., increased activity of the proteins that cause p53 to be recycled).

[D] INSERTION SEQUENCE

Insertion element *(also known as an **IS**, an **insertion sequence element**, or an **IS element**)* is a short DNA sequence that acts as a simple transposable element. Insertion sequences have two major characteristics: they are small relative to other transposable elements (generally around 700 to 2500 bp in length) and only code for proteins implicated in the transpositionactivity (they are thus different from other transposons, which also carry accessory genes such as antibiotic resistance genes).

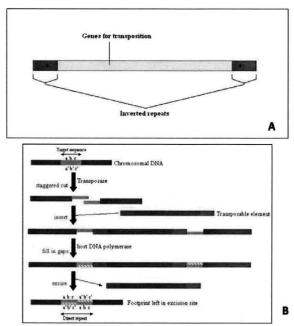

Figure 20: Insartion sequence

Web ref: 17

These proteins are usually the transposase which catalyses the enzymatic reaction allowing the IS to move, and also one regulatory protein which either stimulates or inhibits the transposition activity. The coding region in an insertion sequence is usually flanked by inverted repeats. For example, the well-known IS*911* (1250 bp) is flanked by two 36bp inverted repeat extremities and the coding region has two genes partially overlapping *orfA* and *orfAB*, coding the transposase (OrfAB) and a regulatory protein (OrfA).

A particular insertion sequence may be named according to the form IS*n*, where *n* is a number (e.g. IS*1*, IS*2*, IS*3*, IS*10*, IS*50*, IS*911*, IS*26* etc.); this is not the only naming scheme used, however. Although insertion sequences are usually discussed in the context of prokaryotic genome, certain eukaryotic DNA sequences belonging to the family of Tc1/*mariner* transposable elements may be considered to be, insertion sequences.

An addition to occurring autonomously, insertion sequences may also occur as parts of composite transposons; in a composite transposon, two insertion sequences flank one or more accessory genes, such as an antibiotic resistance gene (e.g. Tn*10*, Tn*5*). Nevertheless, there exist another sort of transposons, called unit transposons, that do not carry insertion sequences at their extremities (e.g. Tn*7*).

A complex transposon does not rely on flanking insertion sequences for resolvase. The resolvase is part of the tns genome and cuts at flanking inverted repeats.

[E] TRANSPOSABLE ELEMENTS & AC-DS SYSTEM

A **transposable element** (**TE** or **transposon**) is a DNA sequence that can change its position within a genome, sometimes creating or reversing mutations and altering the cell's genetic identity and genome size. Transposition often results in duplication of the same genetic material. Barbara McClintock's discovery of these **jumping genes** earned her a Nobel Prize in 1983. Transposable elements make up a large fraction of the genome and are responsible for much of the mass of DNA in a eukaryotic cell. It has been shown that TEs are important in genome function and evolution. In *Oxytricha*, which has a unique genetic system, these elements play a critical role in development. Transposons are also very useful to researchers as a means to alter DNA inside a living organism.

Types of Transposable Elements

The maize *Ac* transposable element is only one of several types found in Nature. Transposable elements can be divided into two major classes based on method of transposition:

[1]   **Retrotransposons** (class 1)

→ Use reverse transposase to make RNA intermediate for transposition.
→ Encode an integrase and reverse transcriptase for transposition.
→ Found in viruses.

[2]   **Transposons** (class 2)

→ DNA fragments transpose directly from DNA segment to DNA segment: Producing a DNA copy that transposes (replicative transposition). Or, cut/paste into a new locus (conservative transposition).
→ Encode a transposase for transposition.
→ Can carry additional genes.
→ Found in eukaryotes and prokaryotes.

The *Ac - Ds* system in maize (*Zea*): genetics of "jumping genes"

In the presence of an **Activator** (*Ac*) element, a **Dissociator** (*Ds*) element can cause chromosome breaks. The genetic marker for such breaks in this example is expression at the *W* locus: the dominant *W* allele produces a mutant White phenotype, and the recessive W^+ allele produces a wild-type dark phenotype. In the above example, the plant is a W / W^+ heterozygote: the other chromosome set is not shown.

(**a**) In the absence of *Ac* element, *Ds* is not transposable and chromosome breaks do not occur. The *W* allele is expressed as expected.

(b) In the presence of an *Ac* element, the **Ds** element can be transposed ("jump") elsewhere in the chromosome. This transposition can cause a chromosome break at the point of insertion, so that the **Ds** element and any downstream loci are lost. In a W / W^+ heterozygote, loss of the W allele would allow expression of the alternative W^+ allele, producing an unexpected dark phenotype

(c) *Ds* transposition can also cause the element to jump into the middle of the W allele, disrupting its normal function and again allowing expression of the alternative W^+ allele. Still under the influence of *Ac*, *Ds* may subsequently jump back out of W, restoring function of W. The plant will then be a mosaic of cell patches, expressing either W or W^+.

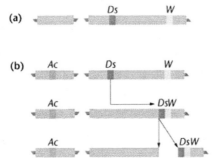

Figure 21: Effect of Transposition involving Ac-Ds system in Maize. (a) In absence of Ac, Ds is not transposable- wild type expression of W occurs; (b) When Ac is present, Ds may be transposed- chromosome breaks and fragment is lost expression of W ceases, producing mutant effect.

Web ref: 18

[F] HARDY WEINBERG GENETIC EQUILIBRIUM

One of the most important principles of **population genetics**, the study of the genetic composition of and differences in populations, is the Hardy-Weinberg equilibrium principle. Also described as **genetic equilibrium**, this principle gives the genetic parameters for a population that is not evolving. In such a population, genetic variation and natural selection do not occur and the population does not experience changes in genotype and allele frequencies from generation to generation.

Hardy-Weinberg Principle

The Hardy-Weinberg principle was developed by the mathematician Godfrey Hardy and physician Wilhelm Weinberg in the early 1900's. They constructed a model for predicting genotype and allele frequencies in a non-evolving population. This model is based on five main assumptions or conditions that must be met in order for a population to exist in genetic equilibrium. **These five main conditions are as follows:**

1. **Mutations** must **not** occur to introduce new alleles to the population.
2. **No gene flow** can occur to increase variability in the gene pool.
3. A very **large population** size is required to ensure allele frequency is not changed through genetic drift.
4. **Mating** must be random in the population.
5. **Natural selection** must **not** occur to alter gene frequencies.

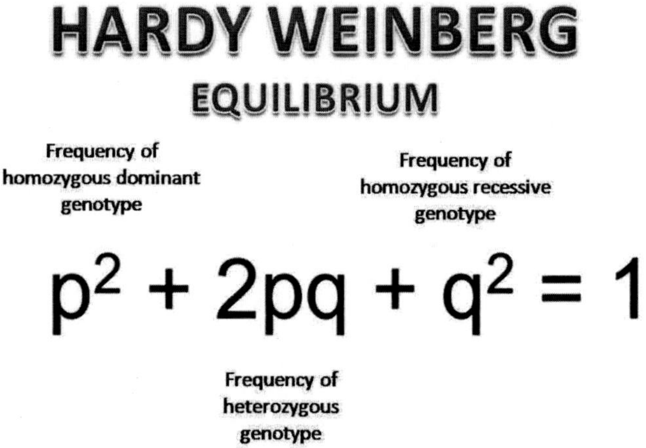

HARDY WEINBERG
EQUILIBRIUM

Frequency of
homozygous dominant
genotype

Frequency of
homozygous recessive
genotype

$$p^2 + 2pq + q^2 = 1$$

Frequency of
heterozygous
genotype

Figure 22: Hardy-Weinberg equilibrium

The conditions required for genetic equilibrium are idealized as we don't see them occurring all at once in nature. As such, evolution does happen in populations. Based on the idealized conditions, Hardy and Weinberg developed an equation for predicting genetic outcomes in a non-evolving population over time.

This equation, $p^2 + 2pq + q^2 = 1$, is also known as the **Hardy-Weinberg equilibrium equation**.

It is useful for comparing changes in genotype frequencies in a population with the expected outcomes of a population at genetic equilibrium. In this equation, p^2 represents the predicted frequency of homozygous dominant individuals in a population, $2pq$ represents the predicted frequency of heterozygous individuals, and q^2 represents the predicted frequency of homozygous recessive individuals. In the development of this equation, Hardy and Weinberg extended established Mendelian genetics principles of inheritance to population genetics.

6.   Genetics problems

Aim: Solve the following genetic problems.

[A] MENDELIAN GENETICS

Example 1:

In dogs, wire hair (S) is dominant to smooth (s). In a cross of a homozygous wire haired dog with a smooth-haired dog, what will be the phenotype of the F_1 generation? What would be the genotype? What would be the ratio of wire-haired to smooth-haired dogs in the F_2 generation?

Ans. The genotype of homozygous wire-haired dog is SS and the genotype of homozygous smooth-haired dog is ss.

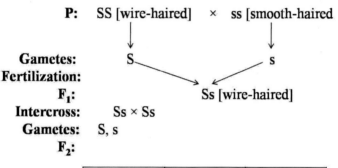

| **P:** | SS [wire-haired] | × | ss [smooth-haired |

Gametes:
Fertilization:
F_1:   Ss [wire-haired]
Intercross: Ss × Ss
Gametes: S, s
F_2:

	S	**s**
S	SS wire-haired	Ss wire-haired
s	Ss wire-haired	ss smooth-haired

Phenotype: Wire-haired : Smooth-haired
Phenotypic ratio: 3 : 1

Example 2: On the basis of Mendel's observations, predict the results from the following crosses with Peas: [a] A tall (dominant and homozygous) variety crossed with a dwarf variety; [b] The progeny of [a] self-fertilized; [c] The progeny from [a] crossed with the original tall parent; [d] the progeny of [a] crossed with the original dwarf parent.

Example 3: An intercross between two pure peas plant yellow- smooth and green- rough is made. If they have two independently assorting genes- one controlling seed color and another controlling seed texture. Suppose yellow (Y) color is dominant to green color (y) and smooth surface (R) is dominant to rough surface (r). [a] Predict the phenotypic outcome of progeny. [b] If the progeny are inter crossed, what phenotype will appear in the F_2 and in what proportions?

Ans.

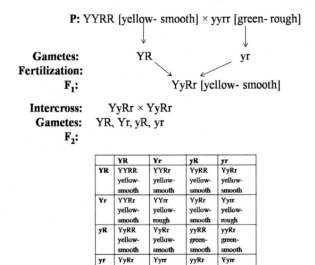

P: YYRR [yellow- smooth] × yyrr [green- rough]

Gametes: YR   yr
Fertilization:
F_1:   YyRr [yellow- smooth]

Intercross: YyRr × YyRr
Gametes: YR, Yr, yR, yr
F_2:

	YR	Yr	yR	yr
YR	YYRR yellow-smooth	YYRr yellow-smooth	YyRR yellow-smooth	YyRr yellow-smooth
Yr	YYRr yellow-smooth	YYrr yellow-rough	YyRr yellow-smooth	Yyrr yellow-rough
yR	YyRR yellow-smooth	YyRr yellow-smooth	yyRR green-smooth	yyRr green-smooth
yr	YyRr yellow-smooth	Yyrr yellow-rough	yyRr green-smooth	yyrr green-rough

Phenotype: yellow-smooth:yellow-rough:green-smooth : green-rough
Phenotypic ratio: 9: 3: 3: 1

Example 4: In Pigeons, a dominant allel (C) causes a chekered pattern in the feathers; its recessive allel (c) produces a plain pattern. Father coloration is controlled by independent assorting gene, the dominant allel (B) produces red feathers and the recessive allel (b) produces brown feathers. Birds from a true breeding chekered, red variety are crossed to birds from a true breeding plain, brown variety [a] predict the phenotype of their progeny. [b] If these progeny are inter crossed, what phenotype will appear in the F$_2$ and in what proportions?

Example 5: An intercross between peas plants heterozygous for three independently assorting genes- controlling plant height, seed color and controlling seed texture. Suppose tall (T) is dominant to dwarf, yellow (Y) color is dominant to green color (y) and smooth surface (R) is dominant to rough surface (r). [a] Predict the phenotypic outcome of progeny by the forked-line method.

Ans.

TtYyRr × TtYyRr

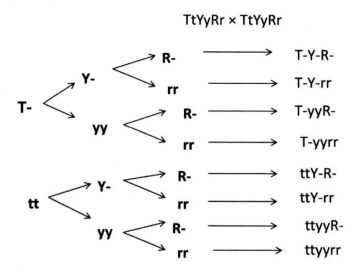

Example 6: An intercross between peas plants heterozygous for three independently assorting genes- controlling plant height, flower color and controlling flower origin on plant. Suppose tall (T) is dominant to dwarf, violet (W) color is dominant to white color (w) and axial flower (A) is dominant to terminal flower (a). [a] Predict the phenotypic outcome of progeny by the forked-line method.

Example 7: A plant heterozygous for three independently assorting genes, AaBbCc is self fertilized. Among the offspring predict the frequency of [a] AABBCC individuals [b] aabbcc individuals [c] individuals that are either AABBCC OR aabbcc [d] AaBbCc individuals [e] individuals that are not heterozygous for all three genes.

Ans.

Because the genes assort independently, we can analyze them one at a time to obtain the answers to each of the questions.

[a] When Aa individuals are selfed, ¼ of the offspring will be AA; likewise for the B and C genes, ¼ of the individuals will be BB and ¼ will be CC

Then, by applying the multiplicative rule of probability to the joint occurrence of three independent events, we can calculate the frequency [that is the probability] of AABBCC offspring as ¼ × ¼ × ¼ = 1/64.

[b] The frequency of aabbcc individuals can be obtained using similar reasoning. For each gene the frequency of recessive homozygous among the offspring is ¼ .

Thus the frequency of aabbcc individuals is ¼ × ¼ × ¼ = 1/64.

[c] To obtain the frequency of individuals that are either AABBCC OR aabbcc, we apply the additive rule of probability and sum of the result of [a] and [b]:

1/64 + 1/64 = 1/32.

[d]   To obtain the frequency of individuals that are triple heterozygotes, again we use the multiplicative rule of probability. For each gene, the frequency of heterozygous offspring is ½ .

Thus the frequency of triple heterozygotes is:

½ × ½ × ½ = 1/8.

[e] offspring that are not heterozygous for all three genes occur with a frequency that is:

1- 1/8= 7/8.

Example 8: How many different kinds of F_1 gametes, F_2 genotypes and F_2 phenotypes would be expected from the following crosses: [a] AA × aa [b] AABB × aabb [c] AABBCC × aabbcc [d] What general formulas are suggested by these answers?

41

Ans.

No.	Cross	F₁ gametes	F₂ genotypes	F₂ phenotypes
[a]	AA × aa	2	3	2
[b]	AABB × aabb	2×2= 4	3×3= 9	2×2= 4
[c]	AABBCC × aabbcc	2×2×2= 8	3×3×3= 27	2×2×2= 8
[d]	general formulas [where n= No. of genes]	2^n	3^n	2^n

Example 9: In a family with three children, what is the probability that two are boys and one is a girl?

Ans. To answer this question, we must apply the theory of binomial probabilities.

For any one child, the probability that it is a boy is ½ and the probability that it is a girl is ½.

Each child is produced independently.

Thus, the probability of two boys and one girl is:

$$\left[\frac{n!}{x!y!}\right](p)^x (q)^y$$

where, n= total no. of children, x= no. of boys, y= no. of girls, p= the probability of child became a boy and q= the probability of child became a girl.

Here, n=3, x=2, y=1, p= ½ and q= ½

So, $\left[\frac{n!}{x!y!}\right](p)^x (q)^y$

$= \left[\frac{3!}{2!1!}\right](½)^2(½)^1$

$= \left[\frac{3×2×1}{2×1×1}\right](½)(½)(½)$

$= \frac{3}{8}$

→ In a family with three children, the probability that two boys and one girl is $\frac{3}{8}$.

Example 10: In a family with six children, what is the probability that two are boys and four are girl?

Ans. 15/64.

Example 11: In a family with six children, what is the probability that one is a boy and five are girl?

Ans. 6/64.

[B] THE HARDY-WEINBERG PRINCIPLE

The **Hardy–Weinberg principle**, also known as the **Hardy–Weinberg equilibrium, model, theorem**, or **law**, states that allele and genotype frequencies in a population will remain constant from generation to generation in the absence of other evolutionary influences. These influences include *mate choice, mutation, selection, genetic drift, gene flow* and *meiotic drive.* Because one or more of these influences are typically present in real populations, the Hardy–Weinberg principle describes an ideal condition against which the effects of these influences can be analyzed.

In the simplest case of a single locus with two alleles denoted A and a with frequencies $f(A)$ = p and $f(a)$ = q, respectively, the expected genotype frequencies are $f(AA)$ = p^2 for the AA homozygotes, $f(aa)$ = q^2 for the aa homozygotes, and $f(Aa)$ = 2pq for the heterozygotes.

The genotype proportions p^2, $2pq$, and q^2 are called the Hardy–Weinberg proportions. Note that the sum of all genotype frequencies of this case is the binomial expansion of the square of the sum of p and q, and such a sum, as it represents the total of all possibilities, must be equal to 1.

Therefore $(p + q)^2 = p^2 + 2pq + q^2 = 1$. A solution of this equation is q = 1 − p.

Population geneticists refer this as **the Hardy–Weinberg genotype frequency**

$[(p + q)^2 = p^2 + 2pq + q^2]$.

This mathematical relationship between allele frequencies and genotype frequencies given by G.H. Hardy and Wilhelm Weinberg called **The Hardy-Weinberg principle.**

Example 12: In the U.S.A., the incidence of the recessive metabolic disorder phenylketonuria (PKU) is about 0.0001. calculate the frequency of the mutant allele that causes PKU?

Ans. First, count the different types of alleles, mutant and normal, that are present in the population. Because, heterozygotes and normal homozygous are phenotypically indistinguishable. So, apply the Hardy–Weinberg principle in reverse to estimate the mutant allele frequency.

The incidence of PKU= 0.0001, represents the frequency of mutant homozygotes in the population.

43

Under the assumption of random mating, these individuals should occur with a frequency equal to the square of the mutant allele frequency. Denoting this allele frequency by q,

$q^2 = 0.0001$

$\therefore q = \sqrt{0.0001}$

$\therefore q = 0.01$

$\therefore$ 1% of the allele in the population are estimate to be mutant

$P + q = 1$

$\therefore P = 1 - q = 1 - 0.01 = 0.99$

$\therefore P = 0.99$

The frequency of people in the population who are heterozygous carriers of the mutant allele:

Carrier frequency = $2pq = 2(0.99)(0.01) = 0.0198$

$\therefore$ **approximately 2% of the population are predicted to be carriers.**

Example 13: An X-linked gene that controls color vision, the frequency of allele for normal vision (C) is p = 0.88 and the frequency of the allele for color blindness (c) is q = 0.12. under the assumptions of random mating and equal allele frequencies in the male is given below.

Sex	Genotype	Frequency	Phenotype
Male	C	p = 0.88	Normal vision
	c	q = 0.12	Color blind

Predict the frequency of the genotypes in female.

Ans. As color blindness is a X-linked homozygous recessive disease, we can predict the frequency of the genotypes in female as given below:

Sex	Genotype	Frequency	Phenotype
Female	CC	$p^2 = (0.88)^2 = 0.77$	Normal vision
	Cc	$2pq = 2(0.88)(0.12) = 0.21$	Normal vision
	cc	$q^2 = (0.12)^2 = 0.02$	Color blind

Example 14: The A-B-O blood types are determined by three alleles I^A, I^B and i. if the frequencies of these are p, q and r respectively, then predict the frequencies of the 6 different genotypes in the A-B-O blood typing system are obtained by expanding the trinomial $(p+q+r)^2 = p^2 + q^2 + r^2 + 2pq + 2qr + 2pr$.

44

Blood type	Genotype	Frequency
A	$I^A I^A$	p^2
	$I^A i$	$2pr$
B	$I^B I^B$	q^2
	$I^B i$	$2qr$
AB	$I^A I^B$	$2pq$
O	ii	r^2

Where, p+ q + r = 1. If the frequency of A gene for homozygote is p = 0.45 and the frequency of B gene for homozygote is q = 0.43, Predict the allele frequencies (i) for AB blood type and (ii) for O blood type.

Ans.

For that, p+ q + r = 1,

∴ 0.45+0.43+r = 1

∴ 0.88+ r = 1

∴ r = 1- 0.88 = 0.12

(i) **The allele frequency for AB blood type:**

The allele frequency for AB blood typ is $2pq = 2(0.45)(0.43) = 0.387$

(ii) **The allele frequency for O blood type:**

The allele frequency for O blood type is $r^2 = (0.12)^2 = 0.0144$

[C] GENETIC MAPPING

Example 15:

Sample Problem: Genetic map

Given the crossover frequency of each of the genes on the chart, construct a chromosome map.

Gene	Frequency of Crossover
A-C	30%
B-C	45%
B-D	40%
A-D	25%

Step 1: Start with the genes that are the farthest apart first: B and C are 45 map units apart and would be placed far apart.

B -------------------------------------- 45% --C

Step 2: Solve it like a puzzle, using a pencil to determine the positions of the other genes.

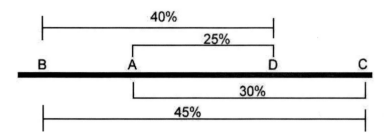

Step 3: Subtraction will be necessary to determine the final distances between each gene.

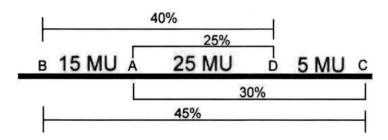

Example 16: In Drosophila, bar shaped eyes (B), scalloped wings (S), Crossveinless wings (W), and Eye Color (C) are located on the X chromosome. The recombination frequency of each gene is indicated on the table. Construct a chromosome map.

Gene	Frequency of Crossover
W-B	2.5%
W-C	3.0%
B-C	5.5%
B-S	5.5%
W-S	8.0%
C-S	11.0%

Example 17: The following chart shows the crossover frequencies for genes on an autosome of the Armor Plated Squirtlesaur. Construct a chromosome map.

Gene	Frequency of Crossover
P-Q	5%
P-R	8%
P-S	12%
Q-R	13%
Q-S	17%

Example 18: Construct a map given the following data.

Gene	Frequency of Crossover
A-B	24%
A-C	8%
C-D	2%
A-F	16%
F-B	8%
D-F	6%

47

7. REFERENCES

Gupta P K (2005) Genetics, Prentice-Hall of India Pvt Ltd., New Delhi (3rd Edition-EEE).

Verma P S and Agarwal (2006) Cell Biology, Genetics, Molecular Biology, Evolution and Ecology. S Chand & Company Ltd., New Delhi (1st Multicolour Edition-Reprint).

Web Refrences:

Web ref-1 https://www.researchgate.net/figure/Plastid-and-its-various-types-with-their-respective-organelle-function_fig1_257429002

Web ref- 2 http://www.qsstudy.com/biology/plastid-definition-functions

Web ref-3
https://ro.wikipedia.org/wiki/Mitocondrie#/media/File:Diagram_of_a_human_mitochondrion.png

Web ref-4 https://www.kullabs.com/classes/subjects/units/lessons/notes/note-detail/1624

Web ref: 5 http://physiologyplus.com/golgi-apparatus/

Web ref: 6 https://prezi.com/er3xv8zo8pe8/animal-cells/

Web ref: 7 https://courses.lumenlearning.com/boundless-microbiology/chapter/other-eukaryotic-components/

Web ref: 8 http://www.bio.miami.edu/dana/151/gofigure/151F12_mitosis.pdf

Web ref: 9 https://www.cliffsnotes.com/study-guides/biology/plant-biology/cell-division/sexualreproduction-meiosis

Web ref: 10 http://www.qsstudy.com/biology/diakinesis-stage-meiosis-plants

Web ref: 11 http://web.as.uky.edu/biology/faculty/kellum/bio315/Lab%204A,%204B%20Exercise-Spring%202015.pdf & http://dnaofbioscience.blogspot.com/2016/05/what-is-gaint-chromosomes.html

Web ref: 12 http://www.bio.miami.edu/dana/151/gofigure/151F12_mitosis.pdf

Web ref: 13
https://biocyclopedia.com/index/genetics/maternal_effects_and_cytoplasmic_inheritance/male_sterility_in_plants.php

Web ref: 14 https://www.ncbi.nlm.nih.gov/pmc/articles/PMC1213516/pdf/323.pdf

Web ref: 15 http://biology.kenyon.edu/courses/biol114/KH_lecture_images/cancer/cancer.html

Web ref: 16 https://www.khanacademy.org/science/biology/cellular-molecular-biology/stem-cells-and-cancer/a/cancer

Web ref- 17 https://commons.wikimedia.org/wiki/File:Insertion_sequence.jpg

web ref-18 http://www.biologydiscussion.com/biotechnology/transposons-definition-and-types-with-diagram/17769

web ref-19 http://selfstudy.co/sp/medical/zoology/cytology/cell-and-protoplasm-theory

Web- http://www.sivabio.50webs.com/plastids.htm

web- https://www.medicalnewstoday.com/articles/320875.php

web- http://www.biologydiscussion.com/plants/cell-wall/cell-wall-meaning-function-and-structure-with-diagram/70471

web- http://www.biozoomer.com/2016/12/golgi-apparatus-golgi-body-structure.html

web- http://www.biologydiscussion.com/dna/2-main-types-of-organellar-dna-cell-biology/38574

web- http://www.yourarticlelibrary.com/biology/chromosome-mapping-notes-on-chromosome-mapping-biology/6599

web- https://www.biologycorner.com/worksheets/genetic_maps.html

web- https://www.chegg.com/homework-help/questions-and-answers/cancer-fundamentally-disease-uncontrolled-cell-division--e-increased-cell-cycle-progressio-q11908653

WEB- https://www.ebi.ac.uk/interpro/potm/2006_12/Page1.htm

web- https://commons.wikimedia.org/wiki/File:Insertion_sequence.jpg

web- http://www.mun.ca/biology/scarr/Ac-Ds_system_genetics.htm

web- https://www.thoughtco.com/hardy-weinberg-equilibrium-definition-4157822